Pharmacokinetics of enantiomers of profens

Master of pharmacy

Dr.G.SANDHYARANI

M.PHARM,Ph.d

Pharmacokinetics of enantiomers of profens

Enantiomers are optically active compounds with one or more chiralcenters and have identical physicochemical properties except the rotation of plane polarized light. Approximately 1 in 4 therapeutic agents are marketed as racemates; the individual enantiomers frequently differ in both their pharmacokinetic and pharmacodynamic profiles. These differences result when the drug molecule has an asymmetric interaction with a receptor, a transport protein or a metabolizing enzyme. The use of single enantiomer has a number of potential clinical advantages including an improved therapeutic and pharmacological profile, a reduction in complex drug interactions and simplified pharmacokinetics (Hutt *et al* 1996)

Drugs given as racemic mixtures have the therapeutic activity mainly in one of the enantiomers. The other enantiomer may be pharmacologically inert or toxic. Almost all profens are enantioselective in their action and disposition.

The 2-Aryl propionic acid (2-APA) derivatives are currently an important group of NSAIDS. A common structural feature of 2-APA NSAIDS is a sp^3–hybridized tetrahedral chiral carbon atom within the propionic acid side chain moity with the S-(+) enantiomer possessing most of the beneficial anti-inflammatory activity (Davies 1998). The 2-APA derivatives under go invivo stereo inversion from R-(-) enantiomer to S-(+) enantiomer and the rate of inversion varies with species and substrate (Brune *et al* 1991). The stereo chemistry of these acids influences their disposition, metabolism, plasma protein binding and elimination.

Metabolism and inversion mechanism

The metabolism of profens involves the stereo specific inversion of the inactive R-(-) enantiomer to active S-(+) enantiomer together with other potentially stereo selective conjugates and oxidative pathways (Bobb *et al* 1979, Goto *et al* 1982). The enantioselective inversion of profens involves the formation of a Coenzyme. A (CoA) thioester followed by epimerization and finally hydrolysis to regenerate the free acids. Inversion requires CoA and

ATPase as cofactors (Hall *et al* 1994, Knihiniki *et al* 1991, Chen *et al* 1990). It was found that liver is the most important site for the inversion (Knihiniki *et al* 1989).

The inversion of the ibuprofen involves 3 steps they are 1) thioesterification of R-ibuprofen to R-ibuprofen Co. A via an adenylate intermediate catalyzed by the long chain acyl-Co.A synthetase. 2) Epimerization of the R-ibuprofenyl Co. A thioester catalyzed by the 2-aryl propionyl Co. A epimerase (Sheih et al 1993). 3) Non-stereo selective hydrolysis of the thioestes (Tracy et al 1992).

Thompson et al (1998) investigations show that *Verticillium lecanii*, a fungus, inverted the chirality of ibuprofen, indoprofen, suprofen, flurbiprofen and fenoprofen from the R-enantiomer to S-enantiomer except ketoprofen in which the inversion was reverse.

The profens are extensively metabolized by oxidation, reduction and conjugation and excreted in urine and feces (Caldwell et al 1988). R-(-) and S-(+) enantiomers of the racemate form the same metabolites but the R-(-) enantiomer inverted into S-(+) enantiomer (Hucker et al P65). Tiaprofenic acid metabolized via oxidation and reduction. Approximately 60-70% of the administered dose eliminated as ester of glucuronide conjugate in urine (Jamali et al 1985).

Arima reported that the formation of acyl glucoronide and glucoside metabolites was dose dependent and time dependent when S-(+) enantiomer, R-(-) enantiomer and RS (+/-) pranoprofen were administered orally (Arima et al 1990, Arima et al 1990). The 2-phenyl propionic acid metabolism involves only glucoronic acid conjugation and under goes chiral inversion (Dixon et al 1997, Hutt et al 1983). Major metabolites of flurbiprofen are 4-hydroxy, 3-hydroxy-4-methoxy and 3, 4-dihydroxy derivatives (Risdal et al 1978). Benoxaprofen, a weak cyclo-oxygenase inhibitor under goes no biotransformation except inversion of R-(-) enantiomer and conjugation (Chatfield et al 1978).

Loxoprofen is a 2-rayl propionic acid derivative with two chiral centers administered as a prodrug. The major metabolites of loxoprofen are cis and trans alcohols, ∝-hydroxy ketone and three diols. The trans-alcohol is optically active with configuration 2S, 1R, 2S and potent inhibitor of prostaglandin synthetase (Naruto *et al* 1984).The cisalcohol under goes irreversible optical inversion to trans-alcohol (Nagashina *et al* 1984). Naproxen follows non-linear pharmacokinetics and metabolism involves the glucoronide conjugation and 6-demethylation followed by glucuronidation (Runkel *et al* 1973). Cicloprofen is excreted in urine mainly in the form of 7-hydroxy derivative along with other three-minor metabolites. The S-(+)-enantiomer of cicloprofen was hydroxylated and excreted at faster rate than its antipode (Lan *et al* 1978, Dean *et al* 1977).

Metabolic chiral inversion *invitro*

Investigation of the chiral inversion of profens with hepatic sub cellular fractions was not very successful. No inversion could be detected for flurbiprofen, naproxen, suprofen, ibuprofen, with CoA, ATP and Mg^{2+} fortified fraction (Mayer *et al* 1988). Rat liver homogenates to which the same factors were added failed to produce the inversion of 2-phenylpropionic acid (Nakamura *et al* 1987). Only marginally significant inversion was found for benoxaprofen (Simmonds *et al* 1980)

The isolated perfused rat liver preparation was the first biological model which allowed the demonstration of the metabolic chiral inversion of ibuprofen *in vitro* (Cox *et al* 1985). Using the isolated perfused rat liver, they showed that the rate and extent of hepatic extraction of ibuprofen decreases with increasing amounts of albumin present in the perfusate (Jenneret *et al* 1990).

Jamali *et al.* (1988) suggested that the inversion may take place in the gastrointestinal tract. The hypothesis of presystemic chiral inversion is supported by the study of Simmonds *et al.* (1980) which showed that (R)-benaxoprofen is inverted in everted gut preparations to its (S)-antipode. However, Nakamura and Yamaguchi (1987) found for 2-phenylpropionic acid only a weak isomerization activity of the small rat intestine compared to that in liver or kidney slices.

An important step towards the identification of the enzyme involved in the metabolic inversion was made by Knights and Meffin (1988). Recirculating a solution of rac-fenoprofen through a column containing hepatic microsomal acyl-CoA synthetase bound to Matrix Gel Red A, they observed that the concentration of (S)-fenoprofen remained constant whereas (R)-fenoprofen disappeared from the solution with formation of the CoA thioester. The enzyme catalyzing stereo selectively the formation of (R)-fenoprofenoyl CoA has been identified as the microsomal long-chain acyl-CoA synthetase.

Hamman *et al* (1997) characterized the regio-and stereo selective metabolism of ibuprofen by human cytochrome P450 liver microsomes. The rate of formation of both the 2-and 3-hydroxy metabolites exhibited monophasic and biphasic substrate concentration dependence for both enantiomers of ibuprofen. The metabolism of ibuprofen indicates that the S-ibuprofen preferred hydroxy ibuprofen, carboxy ibuprofen and ibuprofen glucoronide formation over R-ibuprofen.

Hamdoune etal (1995) reported that the R- (-) enantiomer was glucuronidated faster than the S- (+) enantiomer by liver microsomes of rats and humans and treatment of Phenobarbital enhanced the glucuronidation of R-(-) flurbiprofen than that of the S- antipode.

Metabolic chiral inversion *in vivo*

The enantioselective measurement of the plasma levels of both enantiomers allows to demonstrate the unidirectional nature of the inversion process (Lee *et al* 1985). In man, no inversion has been found for carprofen (Lee *et al* 1986), flunaxoprofen (Pedrazzine *et al* 1988), flurbiprofen (Jamali *et al* 1988) ketoprofen (Sallustio *et al* 1988) and tiaprofenic acid (Singh *et al* 1986).

Factors modulating chiral inversion:

Xiaotao and Hall (1993) showed that pivalic acid inhibits the formation of R-(-)-ibuprofenoyl-CoA in isolated rat hepatocytes; pivalic acid (released from the ampicillin prodrug pivampicillin) is known tobind with intracellular CoA by forming long-lived thioesters, In this case no reduction of inversion was observed. Using the same biological model, they found that the inversion of R-(-)-ibuprofen is inhibited by fatty acids, and they demonstrated that this inhibition is due to a transient depletion of the CoA pool (Mayer *et al* 1995, Muller *et al* 1992). Clofibric acid, a hypolipidemic agent, is known to induce a number of enzymes involved in fatty acid metabolism and causes, in addition, an increase in the total amount of hepatic CoA (Roy *etal* 1996). The effect of clofibrate on the chiral inversion of R-ibuprofen in humans was examined by Scheuerer *et al*. (1996). In their randomized, crossover study, clofibrate pretreatment not only increased the clearance by inversion but also interacts with the oxidative metabolism of ibuprofen enantiomers. Studies in isolated rat hepatocytes showed that xenobiotics interacting with the oxidative metabolic pathways can indirectly influence the extent of inversion (Mayer *et al* 1995). Pretreatment with Phenobarbital, an inducer of cytochrome P450, increased the metabolism by noninversion pathways and consequently reduced the fraction inverted of R-ibuprofen. Not all xenobiotics whose metabolism is CoA-dependent interfere with the chiral inversion of R-ibuprofen. Thus, the presence of benzoic acid or salicylic acid, which undergo glycine conjugation via activation by CoA, dose not affect the chiral inversion of R-ibuprofen in rat hepatocytes (Muller *et al* 1992).

Studies in healthy volunteers:

Avgerinos and Hutt (1990) investigated the plasma disposition of ibuprofen enantiomers following oral administration of the racemic drug in 24 healthy male volunteers. They observed that the plasma elimination of R-(+)-ibuprofen was more rapid than that of the S-enantiomer, 64 % of the total area under the plasma conc. time curves (AUC)was due to the pharmacologically active enantiomer. Furthermore, the influence of dose (200-800mg) on the pharmacokinetic characteristics of the enantiomer of ibuprofen was investigated in 3 subjects. The dose-normalized AUC values and oral clearance showed a dose dependence in the disposition of R-(-)-ibuprofen.

The influence of increasing doses of rac-ibuprofen on the pharmacokinetics of its individual enantiomers was further investigated by Evans *et al*. (1991). In this study, rac-ibuprofen (200, 400, 800 and 1200 mg p.o.) was given on 4 occasions to 4 healthy male volunteers (age 23-28yr, weight 64-78kg). For all 4 doses, the AUC for both the total and unbound fraction in plasma was significantly greater for the S-(+) enantiomer. With increasing rac-ibuprofen dose, there was a less than proportional increase in the total AUC (based on the total plasma concentration) for the enantiomers, which was statistically significant for R-ibuprofen. These results were similar to those observed by other Gabard *et al* (1995). Cheng *et al*. (1994) went a step further and administered rac-ibuprofen and then each enantiomer separately in order to evaluate the potential of enantiomer- enantiomer interaction as well as to determine the rate and the extent of systemic inversion in12 healthy males and concluded that the kinetics of R-(-)-ibuprofen are not altered by the concurrent administration of S-(+)-ibuprofen.

Protein binding:

Profens are bound stereo selectively to serum albumin at 2 sites to different rates. There are two high-affinity binding sites on human serum albumin (HAS) for profens and they are warfarin site (site-I) and benzodiazepine and indole site (Site II) (Fehske *et al* 1981). Drugs bind to site-II in a highly stereo selective manner. The unbound drug is responsible for the pharmacological action; therefore, the differences in protein binding of profens will affect the pharmacodynamic and pharmacokinetic parameters such as volume of distribution (Vd) and clearence. It was observed that (R)-enantiomer of profens highly bound to HAS Carprofen, ibuprofen, indoprofen and phenylpropionic acid bound to serum albumin to higher extent than flurbiprofen and ketoprofen (Oravcova *et al* 1991, Otagiri *eta* 1989).

A study demonstrates that R-(-) and S-(+) ibuprofen had 1 common binding site to HAS and that S-(+) ibuprofen has atleast one other major binding site (Hansen *et al* 1985). Competitive plasma binding between the enantiomers of ibuprofen has been suggested using site I and II specific fluorescent process, the binding of enantiomers was investigated, but no significant difference was found. Utilising astereo specific GC-MS assay with equililed saturable binding with a mean ratio of free concentration of S: R of 1:7 over a concentration range of 2 to 100mg/l. There is an absence of nonlinear protein binding for the enantiomers of ibuprofen at relatively low dose. At low doses (400mg) there were no significant differences between oral and 1.V serum protein binding for R-(-)and S-(+) ibuprofen(Hall *et al* 1993).

Protein binding of ibuprofen enantiomers is stereo selective with greater fraction of S-(+) ibuprofen unbound (Averginos *et al* 1990). In addition, racemic ibuprofen 200 to 1200mg was administered to 4 individuals and the AUC of the total and protein unbound enantiomers were measured .With R-(-) ibuprofen there was a decrease in AUC/dose as the dose was increased. The bound fraction of S-(+) ibuprofen was greater than that of R-(-)-ibuprofen although the AUC unbound versus dose values of the enantiomers remained unchanged (Evans *et al* 1990).

Evans *et al* (1991) determined the unbound fractions of R-G) ibuprofen and S-(+)-ibuprofen applying enantio selective technique which involved the use of radio labeled rainier ibuprofen, this in vitro study showed that the binding of either enantiomer is influenced by the presence of its antipode.

More recently Paliwal *et al* (1993) using more subjects, more data points per subject and separated enantiomers-investigated in vivo the extent of plasma protein binding and estimated the binding Competitive inhibition parameters of R-(-)- and S-(+)- ibuprofen. Stereo selective differences were observed between ibuprofen enantiomers in their binding affinity and to a lesser extent in their competitive inhibitory potential. The intrinsic binding of R-(-)-ibuprofen was greater than S-(+)-ibuprofen and the unbound fraction was greater for S-(+)-enantiomers than for the R-(-)-enantiomer after a given dose of R-(-)-ibuprofen or racemeate.

The S-enantiomer of carprofen and flurbiprofen showed higher AUC and t1/2 and these enantiomers highly bound to proteins (Iwakava et al 1989). Ketoprofen enantiomers showed small differences in protein binding whereas R-2-phenylpropionic acid was eliminated faster than S-antipode leading to greater AUC for S-isomer (Foster et al 1988, Sallustrio et al 1988). In case of indoprofen, the R-enantiomer is more bound and rapidly eliminated than its antipode (Tranmassia et al 1988). Where as repeated administration of flurbiprofen caused accumulation of S-(+) enantiomer (Young et al 1991).Knadler etal(1992) studied the stereo selective disposition of flurbiprofen in ureamic patients and concluded that adjustment of flurbiprofen dosing rate in uramic patients is not in dated on the basis of pharmacokinetics.

Knadler etal (1989) investigated the influence of pathophysiological status on the plasmaprotien binding of flurbiprofen and concluded that the binding of racemic flurbiprofen in elderly and obese volunteers and patients with liver disease was not significantly different from normal subjects; butbindung was less in hypoialbuminic patients and patients with renal impairment.

Berry etal (1989) investigated the enantiomeric ineraction of flurbiprofen in rat and reported that interaction is a result of displace ment from plasma protein binding sites of one enantiomer by the other.

Pharmacokinetics of profens:

2-APA analogues exhibit optical isomerism and because of their stereo selectivity they differ in their pharmacokinetic and pharmacodynamic Properties.

In a preliminary study of the racemate compared with the enantioners {S-(+) 400mg, R-(-)400mg racimic ibuprofen 800mg} the AUC of S-(+)-ibuprofen after racemate was 128 Vs 93.1mg/l 2.hr as compared with S-(+) ibuprofen. The AUC of R-(-) ibuprofen was greater after R-(-)- ibuprofen administration (101 Vs 82.3mg/lts) than after recemic ibuprofen. These initial results suggest the possibility of altered kinetics due to the concurrent administration of the respective optical antipode (Romero *et al* 1991). The bioavailability of S-(+) ibuprofen was independent of dose between 150 and 500mg (Geisslinger *et al* 1990).

Geisslinger *et al* (1995) reported that the R-(-) enantiomer had higher AUC, lower clearenca data and higher C-max values than the S-enantiomer after oral administration of different doses of the ketoprofen racemate.

Menzel *et al* (1994) investigated the effect of dose on the pharmacokinetics of ketoprofen enantiomers in rats in vivo and in hepatoma cells in vitro following administration of the optically pure enantiomers and the racemate of ketoprofen. Independent of the dose administered the fraction inverted in vivo was about 66% and enhanced inversion was observed following racemate as compared to single enantiomer incubation in vitro.

Jamali *et al* (1988) reported dose dependency of flurbiprofen enantiomer pharmacokinetics in the rat by administering different doses of racemic flurbiprofen to rats and statistically significant dose dependent increase in total body clearance and volume of distribution of both the enantiomers.

INTERSPECIES VARIENCE

Some of the 2-APA NSAIDs showed species variation. Benoxaprofen attained high plasma levels in dog, rat, mouse, rabbit between 3and 6 hrs. And in monkey in 1hr. The half-life for racemic drug was 4 hrs in rabbit, 28 hrs in rat and still high in man (Evans *et al* 1978)

Menzel-soglowek *et al* (1992) reported that the inversion of R-flurbiprofen to S-flurbiprofen is different in different species; it was maximum in dog and guinea pig than in rat and Gerbil. Biotransformation of naproxen involves the O-demethylation.6-desmethyl naproxen, a metabolite of naproxen, is found in different concentration in different species. The lowest being observed in humans (Thompson *et al* 1973).

The inversion of R- (-) to S-(+) enantiomer was species dependent in case of 2-phenyl propinoic acid and inversion takes place in rat and rabbit but not in mouse. The formation of metabolite esterglucuronide was enantioselective for the S-(+)-isomer in the rat and mouse but not in rabbit

(Fournel *et al* 1986). There is a variance in loxoprofen metabolism in different species. Loxoprofen monohydroxy metabolites excreted in free form in mice, conjugated form in dogs and ester glucuronide form in monkeys (Tanaka *et al* 1983). Observations show that the inversion of R-(-) to S-(+) cicloprofen is faster in rat than in the monkeys (Lan *et al* 1976).

The stereo selective disposition of profens in laboratory animals does not always correspond to that observed in humans. Thus, the inversion half-life or (R) - to (S)-benaxoprofen in rat is about 2.6 h whereas the conversion in humans is much slower (108 h) (Simmonds *et al* 1980). Flurbiprofen (Jamali *et al* 1988), indoprofen (Buttiononi *et al* 1983), and carprofen (Abas *et al* 1987) are not inverted in either rats or humans. Other compounds, however, which are not inverted in man (namely flunaxoprofen (Pedrazzine *et al* 1988), ketoprofen (Foster *et al* 1987), and tiaprofenic acid (Jamali *et al* 1988) are subject to significant inversion in rats. There are also examples for important interspecies differences. Thus, ketoprofen, which is extensively inverted in the rat, shows less than 10% inversion in rabbit (Abas *et al* 1987). No inversion occurs for 2-phenylpropionic acid in the mouse, in contrast to rat and rabbit (Fournel *et al* 1986). It appears from this discussion that the rate of configurationally inversion depends on both animal species and the nature of substrate.

Bioavailability studies:

The purpose of the study conducted by (Gabard et al 1995) was to define the relationship between the quantity available after administration of pure S-(+)-ibuprofen and the quantity available after its administration as part of the racemate. The bioavailability of S-(+)-ibuprofen (200, 400, 600 mg) was compared with its bioavailability from rac ibuprofen (400, 800, 1200 mg) in 14 male subjects (age 28 ± 3yr, weight 77 ± 7 kg) using a crossover, randomized design. The disposition parameters determined for the ibuprofen enantiomers were in agreement with data from the literature. Just as Cheng et al.(1994) observed that there were no major differences between the pharmacokinetic parameters of S-(+)-ibuprofen administered alone or as part of the racemate. A linear relationship between the AUC and dose was found for S-(+)-ibuprofen when taken alone. After administration of the racemate, dose-normalized AUC values of S-(+)-ibuprofen showed a weak but significant linear association, with a tendency to decrease with increasing dose. A more pronounced decrease in dose-normalized AUC values was found for the R-enantiomer administered as part of the racemate. The mean relative bioavailability of the S-(+)-ibuprofen administered in the racemate was 66%. From this, the authors concluded that the dose of the racemate should be multiplied by a factor of 0.63-0.69 (95% Cl. mean 0.66) in order to find the dose of S-(+)-ibuprofen alone of this enantiomer absorbed.

Absorption characteristics of ibuprofen (oral dosage formulations administered in single doses except where indicated).

Table.

Lee *et al* (1985) observed that 63% of R-(-) ibuprofen inverted to S-(+) enantiomer following separate oral administration of ibuprofen enantiomers and AUC of the S-(+)- ibuprofen was greater than that of R-(-) ibuprofen. It was also found that concurrent administration of the racemic mixture of ibuprofen lowered the AUC of the individual enantiomers. Suri *et al* results also support these findings, that concentration of S-(+) ibuprofen was consistently larger than R-(-) ibuprofen and differed in various pharmacokinetic parameters after oral administration of the racemic ibuprofen (suri *et al* 1997). Cheng *et al* (1994) assessed pharmacokinetics and bioinversion of ibuprofen enantiomers in 12 healthy males. The mean plasma t1/2 of R-(-) ibuprofen was 1.74 hr when infused intravenously as a racemic mixture and was 1.84hr when infused intravenously alone. The mean t1/2 of S-(+) ibuprofen was 1.77hr when dosed as pure enantiomer.. Kinetics of R-(-) ibuprofen were not altered by concurrent administration of S-(+) ibuprofen. In this study there was little or no presytemic inversion of R-(-) ibuprofen to its S-(+) isomer. Also 69% of the intravenous dose of R-(-) ibuprofen lysinate was bioavailability as S-(+) ibuprofen. These results indicate that the bioinversion of R-(-) ibuprofen administered orally is mainly systemic. Because bio inversion of R-(-) ibuprofen is not complete, S-(+) ibuprofen produced higher bioavailability of S-(+) ibuprofen ( ), than either racemic ibuprofen (70.7%) or R-(-) ibuprofen (57.6%). However, bioavailability of R-(-) ibuprofen (83.6%) when dosed alone was not significantly different than when dosed as racimic mixture (80.7%). ( ). A study of ibuprofen (600mg single dose) is 12 healthy volunteers revealed that Terminal half lives were similar for both enantiomers but plasma levels of S-(+)- ibuprofen were higher than those of R-(-) ibuprofen, due to the chiral inversion and differences in distribution and metabolism. Each enantiomer of ibuprofen followed linear pharmacokinetics with no time dependency (Oliary *et al* 1992).

(Walser *et al* 1997) investigated plasma concentration time profiles of both ibuprofen enantiomers in the rat after single oral application of two different crystal forms of S-(+) ibuprofen and racemic ibuprofen in order to optimize blood sampling times in a subsequent sub chronic toxicity study. As the AUC 0-24 h S-(+)- ibuprofen and the AUC 0-24h, R-(-)- ibuprofen after application of commercial and recrystallised crystal species were not different. The crystal form did not exert an influence on the extent of absorption of S-(+) ibuprofen and racemic ibuprofen in the rat. Tia profenic acid (TPA) is a 2-APA NSAID and posses single chiral carbon atom therefore exists as two enantiomers. Erb *et al* reported that the Cmax and AUC of the S-TPA was greater than R-TPA following intra peritoneal administration. This is due to the chiral inversion of the R-TPA to S-TPA (Erb *et al* 1990). Ghezzi *et al* (1998) results show that S-ketoprofen efficiently inhibits the carragenon induced edema and induce the production of inflammatory cytokinine and interleukin-1. The racemic ketoprofen exhibits little stereo selectivity in its pharmacokinetics. Relative bio availability of oral dexketoprofen (12.5 and 25mg, respectively) is similar to that of oral racemic ketoprofen (25 and 50mg, respectively), as measured in all cases by the AUC values for S- (+) –

ketoprofen. After single oral administration of racemic keto profen 50mg as regular release preparation to 8 healthy volunteers, no significant stereo selectivity was found during drug absorption. 3-α hydroxysteroid dehydrogenase is a human liver enzyme responsible for the metabolism of steroid hormones, bile acids, and xenobiotic aromatic hydrocarbons. Isoform of this enzyme possesses activation and inhibition sites. Profens like ketoprofen, fenoprofen, flurbiprofen, ibuprofen, naproxen, and suprofen stimulates the activity of isoform of this enzyme at lower concentration and inhibit at higher concentrations. Whereas R-ibuprofen showed higher stimulation than S-ibuprofen (Yamamoto *et al* 1998)

Jamali *et al* (1988) found that upon iv administration of 10mg/kg of racemic flurbiprofen to male spragues-Dawley rats, the plasma concentrations were consistently higher for S-flurbiprofen than for R-flurbiprofen (AUC=134+/- 3g versus 41+l-9mg l-1h).

Carprofen, a 2-APA derivate is widely used in the veterinary medicine. pharmacokinetics of carprofen in the cat fit a 2-compartment models, with a long +/- A study of pharmacokinetics of carprofens enantiomers in equine plasma and synovial fluid suggests that higher plasma R carprofen concentrations than S-carprofen concentrations throughout the 48-h period. Synovial fluid concentrations of both carprofen enantiomers were significantly lower than plasma concentrations, probably due to high plasma protein binding which could limit transfer through the synovial membrane (Armstrong *et al* 1999).

Sphan *et al (1989)* investigated the influence of probenicid on the pharmacokinetics of carprofen in three healthy volunteers after single period administration of 150mg of RS-(+/-) carprofen. The plasma concentration of S-(+)-carprofen was higher than those of R-(-)-carprofen at most of the sampling points. Probenicid reduced apparent total and renal clearances for both enantiomers. It also reduced the clearance of the carprofen enantiomers to their glucuronides and the renal clearance of the glucuronides. Stereo specific disposition of carprofen, benoxa profen and naproxen were studied in rats after i.v. administration of racemate (11muml/kg) or enantiomer (5.5 mumol/kg). The total clearance of the (R)-enantiomers of Carprofen and flunoxaprofen were significantly greater than those of the (S)-enatiomer. The clearance of (S)-Naproxen was similar to the value for (R)-and (S)-enantiomers of carprofen, flunoxaprofen or naproxen, Biliary excretion of (R)-Carpofen and of its glucoronide were higher than those of the (S)-enantiomer and its glucoronide.

Brocks *et al* (1993) studied pharmacokinetics of the enantiomers of the NSAID pirprofen in male Sprague – Dawly rats after oral and intravenous (iv) doses of the racemate. No significant differences were identified between the enantiomers after oral or iv dosing in $T_{1/2}$, V_d, or ΣXu. The absolute bioavailability of the active S-enantiomer after oral doses was higher than the inactive R-enantiomer.

Clinical Significance:

Racemate is an equimolar mixture of R-(-) and S-(+) enantiomers. In case of profen, racemic mixture contain active S-(+) and inactive R-(-) enantiomers. The contribution of R-enantiomer is more towards side effects, drug interactions and little towards therapeutic effect.

Geisslinger etal (1993) investigated the pharmacodynamic behaviuor of pure enantiomers of R-&S- flurbiprofen in the rat and concluded that R-flurbiprofen which is not an inhibitor of prostaglandin synthesis in vitro had only marginal anti inflammatory activity as defined by the carragenan edema of the rat paw in contrast to the S- enantiomer.Wetcher etal (1993) reported that rac- flurbiprofen is more ulcerogenic than its (S)-enantiomer.In astudy conducted by Bonabello etal (2003) the S(+)-ibuprofen found to be more potent than the racemic formulation and produced less acute gastric damage.

Dexketoprofen trometalal is a water soluble salt of the dextro rotatory enantiomer of the (NSAID) ketoprofen. In humans, the anti-inflammatory potency of dexketoprefen was always equipotent to that demonstrated by twice the dose of ketoprofen. On set of action appeared to be shorter and well tolerated by the patients. (Mauleon etal 1996)

The regulatory authorities must implement regulations which include the instructions to manufacturers to label the character of drugs, mixture of stereo isomers (R/S or cis/trans) with their molar ratios, pharmacokinetics and pharmacodynamics of individual enantiomer. Thalidomide and benoxaprofen incidences show importance of stereoisomerism studies.The different pharmacokinetic behavior of enantiomers appears to be a contributor of the adverse reactions that led to the withdrawal of benoxaprofen from the market in 1982 (De camp *et al* 1989, Marshal *et al* 1985).

Almost all the presently marketed dugs are recemates. Naproxen is the only NSAID available in pure (S)-enaniomer form. (S)-Ibuprofen has been available in Austria since 1994 onwards and (S)-ketoprofen has been recently marketed in Spain. Japan is the first country to implement the regulations for racemic products. Full pharmacology and toxicology are required for each of the enantiomers and for the race mates. The European Economic Community (EEC) report states that "where a mixture of stereo isomers has previously been marketed and it is now proposed to market a product containing only one isomer, full data on this isomer should be provided".

Therefore it is advisable that racemates should be withdrawn from the market as soon as pure individual enantiomer preparations are available at an acceptable price.

Studies on ibuprofen:

Chiral inversion of Ibuprofen:

Ibuprofen is a major nonsteroidal anti-inflammatory drug which undergoes unidirectional chiral inversion to the active (S)-ibuprofen *in vivo* and in various biological *in vitro* preparations (Mayer *et al* 1990, Mayer *et al* 1997). This unidirectional chiral inversion may have potential toxicological significance since the acyl-CoA thioester can modulate lipid metabolism. Thus, the acyl-CoA thioester can inhibit mitochondrial β-oxidation of fatty acids, be incorporated into triglycerides, and lower serum lipid levels. In rat liver homogenates, the formation of (R)-ibuprofenoyl-CoA is dependent on the concentrations of both CoA and (R)-ibuprofen. The implication is that compounds which alter intracellular concentrations of CoA should influence the rate o f chiral inversion of (R)-ibuprofen. Such an interaction has been demonstrated for long-chain fatty acids, which inhibit the inversion of (R)-ibuprofen in rat hepatocytes by competing with (R)-ibuprofen for both CoA and acyl-CoA ligases (Mayer *et al* 1996). In contrast, clofibric acid, a hypolipidemic agent, increased the chiral inversion of (R)-ibuprofen in perfused rat liver and rat hepatocyte suspensions (31). This increased inversion of (R)-ibuprofen in the presence of clofibric acid is probably the result of a shift in intracellular pools of CoA. (Roy *et al* 1996).

Metabolism:

The metabolism of rac-ibuprofen involves three routes: 1) oxidation of the aliphatic side chain, 2) conjugation to an acylglucuronide, and 3) inversion of the R-(-)-to the S-(+)-enantiomer.

Oxidation:

Oxidation is a major metabolic pathway for ibuprofen. In humans following oral administration of rac-ibuprofen, more than 60% of the dose is recovered in urine (in a 24 –h collection period) as 2-hydroxyibuprofen, 3-carboxyibuprofen (Fig.3) and their conjugates (Rudy et al 1990, Rudy). The observed metabolite formation clearance indicates a stereo selectivity in favor of the S-enantiomer and regioselevtivity in favor of 3-carboxyibuprofen. However, these metabolic pathways display a relatively modest stereo selectivity. Thus, the unbound formation clearances of 2-hydroxyibuprofen and 3-carboxyibuprofen exhibit S/R enatioselectivities of 1.1 and 1.7, respectively (Knadler et al 1990, Rudy et al 1995). The oxidized metabolites can subsequently undergo glucuroconjugation. There is evidence that the hydroxylation of the ibuprofen enantiomers is mediated by cytochrome P450 2C9 (CYP2C9). Indeed, Leeman et al. (1993) observed that2-and 3-hydroxylation can be effectively inhibited by sulfaphenazole, a specific inhibitor of this enzyme.

Glucuronidation:

Conjugation of ibuprofen with glucuronic acid (Fig.3) is relatively a minor pathway in humans (< 10% of the dose) . Hayball et al. (1993) suggested that renal insufficiency may alter the pharmacokinetics of 2-arylpropionic acid via "futile cycling" of the acyl glucuronides. Rudy et al. (1995) however, pointed out that such a phenomenon is unlikely to be important in the case of ibuprofen since this compound is eliminated only to a minor extent by glucuronidation.

Thioesterification with CoA;

Tracy *et al*, (1993) compared the formation of ibuprofenyl CoA in both rat and human liver tissue. They found that rat whole liver homogenate is approximately 4 times more efficient at formingIbuprofenyl-CoA than human liver tissue. Investigations in tissue homogenates have shown that the formation of the CoA thioester requires the presence of CoA, ATP and $MgCl_2$ (Knadler *et al* 1990, Tracy1993). Menzel *et al*. (1994) examined more precisely the role of ATP in thioesterfication. Theyfound that adenylate formation prior to thioesterification is the first and stereo selective step of R-ibuprofen inversion.

Epimerisation:

The existence of a "2-arylpropionyl-CoA epimerase" was first demonstrated by

Shieh *et al* (1993). They purified an epimerase to homogeneity from rat liver cytosol and mitochondria and found that both forms are monomeric proteins with a molecular mass of 42kD, but differ in their amino acid composition. Incubations of chemically synthesized CoA thioester of the ibuprofen enantiomers in sub cellular rat liver preparations showed that epimerization is freely reversible from both directions (Knihinicki *et al* 1989,Chen *et al* 1990,Tracy *et al* 1992). Using a purified microsomal 2-arylpropionyl-CoA epimerase, Shieh *et al* (1993) also found a bidirectional epimerization, but showed that the apparent initial rate of epimerization is higher for S-ibuprofen than for the R-enantiomer. The equilibrium constant $K_{eg} = \{R\} / \{S\}$ was estimated to be 1.5. Reichel *et al.*(1993) examined the localization and activity of the epimerase in various tissues of guinea pigs and rats. They isolated and purified a 42 kD epimerase isolated from the cytosolic fraction of rat livers and raised polyclonal antibodies against this enzyme in rabbits. Western blot analysis revealed the presence of the enzyme in all organs of both species. The inversion of R-(-)-ibuprofen and its thioester, however, was found to take place predominantly in the liver and kidney and to some extent in the heart and ileum

Hydrolysis of the CoA thioesters:

The final step in the inversion pathway is the hydrolysis of the epimeric CoA thioesters to release the free arylpropionic acid. Investigations of the hydrolysis of ibuprofenoylCoA with synthetic CoA thioesters of ibuprofen enantiomers demonstrated that this reaction is enzyme-catalysed. The CoA thioesters are stable in aqueous solutions at 37°C, buffered to physiological pH. They are only slowly hydrolyzed when incubated with human plasma. In contrast, rat liver mitochondria and microsomes rapidly hydrolyzed both R- and S-ibuprofenoyl-CoA (Knihiniki *et al* 1991). In rat liver homogenates and mitochondria, the rate of hydrolysis of R-ibuprofenoyl-CoA h as the same as that of the S-enantiomer.. Tracy *et al* (1991) found that the hydrolysis of the acyl-CoA thioesters is inhibited by high concentrations of CoA and ATP in rat liver homogenates. This suggests that the ATP and CoA levels modulate the rate of hydrolysis of ibuprofen-CoA thioesters by a feedback mechanism.

Influence of age on the disposition of ibuprofen enantiomers:

Age can be an important factor affecting the pharmacokinetics of drugs. In the case of profens, the situation is further complicated by the occurrence of stereo selective processes. Two studies have been published which assess the influence of age on stereo selective pharmacokinetics. The first was carried out in infants and the second in elderly subjects.

The pharmacokinetics of ibuprofen enantiomers in 11 infants (7 males and 4 females, age 6-18 months, weight 9.5 $\pm$ 1.6 kg) was investigated following administration of a single oral dose (7.6 $\pm$ 0.3 mg/kg) of the racemate for postoperative analgesia. In infants, the plasma concentrations of the S-(+)-isomer were significantly lower than those of the R-(-)-form, while in adults the S-(+)-isomer predominated. Explanations for this difference include the possibility of an impaired R-to-S inversion and /or higher clearance of the S-isomer in infants. Since plasma concentrations of the active isomer are low in infants (relative to adults), a higher dosage might be required in infants (Rye *et al* 1994).

Using a stable-isotope methodology, Rudy *et al.* (1995) compared the disposition of ibuprofen in young healthy volunteers (n=8, age 28 $\pm$ 4 yr) with that in healthy elderly volunteers (n=14, age 73 $\pm$ 5 yr). The volunteers received a 4-week regimen of 800 mg of rac-ibuprofen every 8 h. The results of this study showed that the fractional inversion of R-(-)-ibuprofen did not change with age and that age had only minimal influence on the pharmacokinetics of the total concentrations of the individual enantiomers after single and multiple dosing. .The study revealed that age per se is associated with a 2-fold increase in the unbound concentration of S-ibuprofen probably due to an impairment of its metabolism. Indeed, in the elderly subjects, a modest but significant reduction in the metabolite formation clearance was noted particularly for S-ibuprofen glucuronide and to a lesser extent for S-hydroxyibuprofen.

Influence of gender on the disposition of ibuprofen enantiomers:

While Rudy *et al*. (1995) were primarily interested in the influence of age on the pharmacokinetics of ibuprofen; their data also indicated that the influence of gender is not statistically significant. The research of knights *et al*. (1995) would appear to confirm this conclusion. Their study was specifically designed to investigate the effect of gender (and oral contraceptive steroids) on the pharmacokinetics of R-(-)-ibuprofen. No significant differences between the groups were found for the values of AUC, oral clearance, half-life and steady-state volume. Similarly, there was no statistically significant difference between the percentage of unbound R-(-)-ibuprofen in pooled plasma in the three groups. Knights *et al*. (1995) concluded from these data that gender and oral contraceptive steroids have little or no effect on the conversion of R-(-)-ibuprofen into the pharmacologically active S-enatiomer.

Influence of Disease State:

Similar conc ratios in synovial fluid and plasma have been reported for ibuprofen in patients with arthritis and chronic knee effusions (Day Ro *et al* 1988). Inpatients with nerve-root compression pain requiring a lumbar the plasma and CSF concentrations of ibuprofen enantiomers have been determined both enantiomers peaked later in the CSF (3hrs) than in the plasma (1.5 hrs) and at lower concentrations. The estimated t1/2 B of R-(-) and S-(+)- ibuprofen user 1.7 and 2.5 has in plasma and 3.9 and 7.9 hrs in the csf respectively. The CSF concentration of both enantiomers was higher than their concurrent free plasma concentrations from 1.5 to 8 hrs (Bannwarth *et al* 1995).

Most patients with renal insufficiency show elevated plasma S-(+)-ibuprofen, higher AUC for S-(+)-ibuprofen and increased AUC.S: R ibuprofen ratios compared with healthy individuals. A reduced clearance.

Future trends

Recemates are equimolar mixture of two enantiomers that differ in their pharmacokinetics and pharmacodynamics. It is well established that only S-(+) enantiomer of profens is a potent inhibitor of prostaglandin synthesis. Almost all profens are administered as racemates, therefore, the R-(-) enantiomer must under go chiral inversion in order to achieve desired anti-inflammatory activity. The enentiomers of the profens and their metabolites are identified using enantioselective analytical methods(Tan 1994), chromatographic resolution technique (Hault 1999).Recent interesting finding is that *Verticillium lecanii,* a fungus, inverted the chirality of the profens and this can be used for the biotechnological production of pure enantiomers of profens. It is desirable to search for more better methods to resolve the racemates to get individual enantiomers and there is need for better understanding of pharmacokinetic ,pharmacodynamic and toxicological properties of individual enantiomers .

References:

Abas A. and Meffin P. J. *J. Pharmacol. Exp. Ther. 240,* 637 (1987).

Acrous L, Khan Az, Gerennan DM, Alam-siddiqi M. the binding of flurbiprofen to plasma protein. J. Pharma. Pharmacol. 1985 Sep:; 37 (9): 644-6.

animals and man. Xenobiotica., (1978),8, 133.

anti inflammatory drugs – IV – ketoprofen disposition.. *Biochem. Pharmacol. 35,* 4153 (1986).

Ariens, E.J., Chirality Sunshine, I., Alto.P.ledd (eds), Marcel Dekker series, Newyork, NY, 1992.

Arima,N.Acyl glucoronidation and glucosidation of pranoprofen a 2-arylpropionic acid derivative in mouse liver and kidney homogenates J.Pharmacobiodyna., (1990) (12),724-32.

Arima, N. Stereoselective acyl glucuronidation and glucosidation of pranoprofen, a 2- aryl propionic acid derivative in mice invivo. J. Pharmacobiodyna, (1990)13(12),733.

Armstrong, S., Tricklebank, P., Lake, A., Frean, S. and Lees, P., Pharmacokinetics of carprofen enantiomers in equine plasma and synovial fluid-a comparision with ketoprofen. J.Vet.Pharmacol. Ther., (1999), 22(3),196-201.

Averginos A. Hutt AJ. Inter individual variability in the enantiomeric disposition of ibuprofen following the oral administration of the racemic drug to healthy volunteers.

Averinos, A., Hutt, A.J. Inter individual variability in the enantiomeric disposition of ibuprofen following the oral administration of the racemic drug to healthy volunteers. Chirality 1990, 2: 249-56.

Berry BW, Jamali F, Enantiomeric interaction of flurbiprofen in the rat. J. Pharma Sci. 1989; 78(8): 632-4.

Bopp R.J., Nash, J.F., Ridolfo, A.S. and Shepard, E.R. Stereoselective inversion of (R) – (-) – benoxaprofen to the (S) – (+)- enantiomer in humans., Drug. Metab. Dispos., (1979),7,356-359.

BrightonUK, 1978, Sep.10.

Brocks, D.R., Liang, W.T. and Jamali, F. Influence of route of administration, on the pharmacokinetics of pirprofen enantiomers of pirprofen enantiomers in the rat. Chirality., (1993),5(2),61-4.

Brune, K., Beck, W.S., Geisslinger, G., Menzel-Soglowek, S and Peskar, B.M.Aspirin like drugs may block pain independently of prostaglandin synthesis inhibition. Experientia., (1991), 47, 257-261.

Buttiononi A., Ferrari M., colombo M. and Ceserani R. Pharmacokinetics of ibuprofen evautionurs in human following oral administration of tablets with different absorption rates. *J. Pharm. Pharmacol. 35* 603(1983).

Caldwell,J., Hutt, A.J. and Fournel-giglex, S. The metabolic chiral inversion and Dispositional Enatioselectivity of the 2-aryl propionic acids and their Biological Consequences. , Biochem. Pharmacol., (1988), 37, 105.

Chatfeild DH.,and Green,J.N., Disposition and metabolism of benoxaprofen in laboratory

Chen, C.S., Chen, T. and Sheih, W.R., Metabolic stereoisomeric inversion of 2-aryl propionic acids on the mechanism of ibuprofen epimerization in rats. Biochem. Biophys. Acta., (1990), 29(1033(1), 1-6.

Cheng, H., Rogers, J.D., Demetriades, J.L., Holland, S.D., Seibold, J.R. and Depuy, E.Pharmacokinetics and bioinversion of ibuprofen enantiomers in humans. Pharm. Res., (1994), 11,824-30.

Chirality 1990; 2: 249-56.

Cox J. W., Cox S.R., van Giessen G. and Ruwart M.J.Ibupreofen stereoprofen hepatic cleartence and distribution in normal and fatty in situ perfused rat liver.J. Pharmacol.Exp. Ther., (1985),232,636.

Davies N.M .Clinical pharmacokinetics of ibuprofen .The first 30 years. Clin. Pharmacokinet., (1998), 34(2), 101-54

Davies,M.N. and Skjodt, M.N.Choosing the right non steroidal anti inflammatory drug for the right patient. Clin.Pharmcokint., (2000), 38(5), 377-92.

Dean,A.V., Lan, S.J., Kripalani, K.J., difazio, L.T. and Schreiber, E.C.,Metabolism of the (+)-,(+/-),- and (-) enantriomers of alpha methyl flourene –2-acetic acid (cicloprofen) in rats. Xenbiotica., (1977),7(9), 549-60.

DeCamp,W.H.The FDA perspective onj the development of stereo isomers.Chirality.,1986 1), 2-6.

Dixon, P.A.F., Caldwell,J. and Smith, R.L., Metabolism of aryl acetic acids .3. Themetabolic fate of diphenyl acetic acid and its variatioa with species and dose. Xenibiotica., (1997), 7, 707.

Erb, K., Brugger, R., Williams, K and Geisslinger, G. ,Stereo selective disposition of Tiaprofenic acid enantiomers in rats. Chirality., (1990), 11, 103-108.

Evans AM Nation RL. Sansom LN. *et al*. The relationship between the Parmacokinetics of Ibuprofen enantiomers and the dose of racemic ibuprofen in humans. Biopharm Drug Dispos 1990; 2: 507-18

Evans AM, Nation RL, Sansom LN, Bochner F, Somogyi AA. Stereoselective plasma protein binding of ibuprofen enantiomers. European Journal of Clinical Pharmacology *36*: 283-290.

Evans AM. Nation RL. Sansom LN. *et al*.Stereoselective plasmaprotien binding of ibuprofen enantiomer. Eur J Clin Pharmacol 1989; 36: 283-90.

Evans, A.M., nation, R.L., Sansom, L.N., Bochner, F., Somogyl, A.A., Br. J. Clin Pharmacol 1991, 31:131-8.

Evans,D., Kitchen, E.A. and Meacock, S.C.R., Brighton symposium on medicinalchemistry

Fehske, K.J., Muller, E.and Woller, The location of drug binding Sites in human Serum Albumin.J., Biochem. Pharmacol., (1981),30,687-92.

Foster RT, Jamali F, Russel AS, Alballa SR. Pharmacokinetics of ketoprofen enantiomers in young and elderly arthritic patients following single and multiple doses. *Journal of pharmaceutical Sciences 77*: 191-195,1988b.

Foster,R.T., Jamali,F., Russel, A.S. and Alballa,S.R.,Pharmacokinetics of ketoprofen enantiomers in healthy subjects following single and multiple doses. J.Pharm.Sci., (1988),

Fournel S. and Caldwell J. Enantio selective disposition of 2-aryl propionic acid nonsteroidal

Fournel,S and Caldwell,J. . The metabolic chiral inversion of 2-phenyl propionic acid in rat, mouse and rabbit., Biochem.Pharmacol., (1986), 1:35, 4153-9.

Gabard, B., Nirnberger, G., Schiel, H., Kikuta, C., Mayer, J.M. Comparison of dexibuprofen administered alone or as part of racemic ibuprofen. Eur.J.Clin Pharmacol. (1995), 48, 505-11.

Gabard, B., Nirnberger, G., Schiel, H., Mascher, H., Kikuta, C., Mayer, J.M.Comparision of Dexibuprofen administered alone or as racemic ibuprofen. Eur J Pharmacokin Biopharm 1993, 21:145-61.

Geisslinger G, Schuster O, Stock KP, *et al*.Pharmacokinetics of S-(+) and R-(-) ibuprofen in volunteers and first clinical experience of S-(+) ibuprofen in rheumatoid arthritis. Eur J Clin Pharmacol 1990. 38, 493-7.

Geisslinger. G., Schuste O. Stock., K.P., *et al*. Pharmacokinetics of S-(+)- and R-(-)-ibuprofen in volunteers and first clinical experience of S-(+)-ibuprofen in rheumatoid arthritis. Eur. J.Clin. Pharmacol. (1990), 38, 493-7.

Ghezzi, P., Marullo, A., Sabbatini, V., Caselli, G. and Bertini, R., ifferentialcontribution of R and S isomersion ketoprofen Anti-inflammatory Activity: role cytokine modulation. J.Pharmacol Exp. Ther., (1998), 287(3), 969-74.

Guisslinger G, Menzel S, Wissel K, Brunck. Pharmacokinctics of kectoprofen enantionurs after different dioses of the racemate. Br J. Clin. Pharmacol., 1995; 40 (1): 73-5.

Hage DS. Noctor TAG. Wainer IW.Characterisation of the protein binding of chiral drugs by high performance affinity chromatography. Interactions of R- anmd S-ibuprofen with human serum albumin. J. Chromatogr 1995; 693: 23-32.

Hall SD. Rudy AC. Knight PM. *et al*. Clin Pharmacol Ther 1993; 53: 393-400.

Hall,S.D. and Quan, X.,The role of co-A in the biotransformation of 2-arylpropionic acids. Chem.Biol.Interact., (1994), 90(3), 235-51.

Hamdoune M, Mounic J, Magd a lou J, Masmoudi J, Goudonnct H, E Scourse A. Charaterization of the in vitro slucuronidation of flurbiprofen enantionurs. Drug metab. Dispos. 1995: 23(3): 343-8.

Hamman, M.A., Thompson, G.A. and Hall, S.D.Regioselective and stereoselective metabolism of ibuprofen by human cytochrome p4502C. Biochem. Pharmacol., (1997), 1.54(1), 33-41.

Hansen T, Day R, Wiliams K, Lee E, Knihinicki R, *et al*. the assay and in vitro binding of the enantiomers of ibuprofen, Clinical and Experimental Pharmacology and Physiology 9(Suppl.): 82-89, 1985.

Hayball. P.J., Nation, R.L., Bochner, F., sansom, L.N., Ahern, M.J., Smith, M.D. The influence of renal function on the enantioselective pgarmacpokinetics and pharmacodynamics of Ketoprofen in patients with rheumatoid arthritis.Br. J. Clin. Pharmacol.l 1993, 36: 185-93.

Hoult,J.R.,Jackson,B.R.,Benicka,E.,Patel,B.K.andHutt,A.J.,Chromatographicresolution,chiroptical characterization and preliminary pharmacological evaluation of the enantiomers of butibufen :a comparision with ibuprofen .pharm.Pharmacol., (1999), 51, 1201-5.

Hucker, H.B., Kwan,K.C. and Duggan,D.E. In: progress in drug metabolism (Bridges, J.W. and Chasseand, L.F eds). Vol:5, p65, Wiley Chichester.

Hutt, A.J. and Caldwell,J., The metabolic chiral inversion of 2-aryl propionic acids – a novel rate with pharmacological consequences. J.Pharm.Pharmacol., (1983)35,693.

Hutt, A.J. and Tan, S.C., Drug chirality and its clinical significance . Drugs., (1996), 52,(suppl.5), 1-12.

ibuprofen and flurbiprofen in rats. Chirality, 1990, 2: 134- 40.

ibuprofen.J.Pharmacol.Exp.Ther., (1991), 259(3), 1133-9.

Itoh,T.,Saura,Y., Tsuda,Y. and Yamada,H.Stereoselectivity and enantiomer-enantiomer interactions in the binding of ibuprofen to human serum albumin. Chirality., (1997),9(7), 643-9.

Iwakawa S, Spahn H, Benet L, Lin E. Stereoselective binding of the glucuronide conjugates of carprofen enatiomers to human serum albumin. *Biochemical Pharmacology 39*: 949-953, 1990.

Iwakawa,S., Spahn, H., Benet,L.Z. and Lin,E.T., Stereoselective disposition of carprofen, uoxaprofen and naproxen is rats. Drug.Metab.Dispos., (1991), 19(5), 853-57.

Iwakawa,S., Suganuma,T., Lee,S.F., Spahn, H. and Benet, L., Direct determinatiopn of diastereomeric carprofen glucuronides in human plasma and urine and preliminary

J.M. Mayer and B. Testa, Pharmacodynamics, pharmacokinetics and toxicity of ibuprofen enantiomers.Drugs of the Future, 22,(1997), pp.1347-1366.

J.M. Mayer, Stereoselective metabolism of anti inflammatory 2- aryl propionates. Acta Pharm. Nord. 2., (1990), pp. 197-215.

J.M. Mayer. Ibuprofen enantiomers and lipid metabolism. J.clin. Pharmacol. 36, (1996), pp.27s-32s.

Jamali F., Berry B. W., Tehrani M.R. and Russell A. .Stereoselective pharmacokinetics of Flurbiprofen in humans and rats. j. Pharm. Sci. *77*, 666(1988).

Jamali F. Pharmaco kinetics of enantiomers of chiral non steroidal anti inflammatory drugs. Eur. J. Drug Metab. Pharmacol. *13*, 1 (1988).

Jamali, F., Berry, B.W., Tehrani, M.R. and Russell, A.S. Stereo selective pharmacokinetics of flurbiprofen in humans and rats. J.Pharm. Sci., (1988), 77(8), 666-9.

Jamali, F., Russel, A.S. and Lehmann, E., Pharmacokinetics of tiaprofenic acid in healthy and arthritic subjects. J.Pharm.Sci., (1985), 74(9), 953-63.

Jeanneret F. Ph.D., Thesis. University of Lausanne, Switzerland., 1990.

Jones ME, Sallustio BC, Purdie YJ, Meffin PJ. Wnatioselective disposition of 2-arylpropionic acid nonsteroidal anti-inflammatory drugs. II. 2-phenylpropionic acid protein binding. *Journal of Pharmacology and Experimental Therapeutics 238*: 288-294, 1986.

Kemmerer J. M., Rubio F. A., McClain R. M. and Koechlin B.A. Stereo specific assay and stereospwcific disposition of racemiccarprofen in rats *J.Pharm. Sci.68*, 1274 1979).

Kinhimnicki, R.D., Day, R.D. and Williams, K.M.Chiral inversion of 2-aryl propionic acid

Knadler MP, Brater DC, Hall SD. Plasma protein binding of flurbiprofen: enantioselectivity andinfluence of pathological status. *Journal of Pharmacology and Experimental Therapeutics 249*: 378-385, 1989.

Knadler, M.P., Hall, S.D.Stareo selective aryl propionyl –coA thio ester formation in vitro. Chirality, 1990, 2: 67-73.

KnadlerMP,Brater DC, Hall SD. Stereo selective disposition of flurbi profen in uracneic patients Br. J. Clin Pharmacol. 1992 Apr; 33 (4): 377-83.

Knights K.M. and meffin p. Enantiospecific formation of fenoprofen co enzyme A thioester in vitro. J biochem. Pharmacol., (1988), 37, 3539.

Knigts KM. McLean CF. Tonkin AL. *et al*. Lack of effect of gender and oral contraceptive steroids on the pharmacokinetics of R-(-) ibuprofen in human. Br J Clin Pharmacol 1995; 40:153-6.

Knihinicki, R.D., Day, R.O., Graham., G.G., Williams, K.M.Stereo selective disposi5tion of

Knihinicki, R.D., Williams, K.M. and Day,R.D. Chiral inversion of 2- aryl pror pionic acd nonsteroidal anti inflammatory drugs-11.In vitro studies of Ibuprofen and flurbiprofen.Biochemicasl pharmacol (1989),15(38), 4389-95.

Lan,S.J., Dean,A.V.,Kripalani,KJ.and Cohen,A.l, Metabolism of alpha methyl flourene 2-acetic acid (cicloprofen) isolation and identification of metabolites from rat urine. Xenbiotica., (1978), 8(2), 121-31.

Lan, S.J., Kripalani, K.J., Dean, A.V., Egli, P., Difazio, L.T. and Schreiber, E.C., Inversion of ptical configuration of alpha methyl flourene-2-acetic acid (cicloprofen) in rats and monkeys. Drug.Metab.Dispos., (1976), 4(4), 330-9.

Lapique, F., Muller, N., Paya, E. and Dubios, N. Protien binding and stereo selectivity of onsteroidal anti inflammatory drugs. Clin.Pharmacok., (1993), 25(2), 115-125.

Lee, E.J.D., Williams, K., Day, R., Graham, G. and Champion, D., Stereoselective disposition of ibuprofen enantiomers in man. Br.J.Clin.pharmacol., (1985), 19, 669-74.

Leeman, T.D., Transon, C., Bonnabry, P., dayer, P.A major role for cytochrome p450 TB(CYP 2C sub family) in the actions of nonsteroidal anti inflammatory drugs. Drugs. Exp. Clin. Res. 1993, 19: 189-95.

Li. G. Treiber G. Maier K. *et al*. Disposition of ibuprofen in patients with cirrhosis. Clin Pharmacokinet 1993; 25: 154-63.

Marshall, E., Guilty plea puts oraflex case to rest. Science. (1985), 229, 1071.

Mayer, J.M., Testa, B., Roy-de Vos, M., Audergon, C., Etter, Interactions between the in vitro metabolism of xenobiotics and fatty acidsaJ.C. Toxicol 1995, 17 (Suppl.) 499-513.

Mckellar, Q.A., Delatour, P. and Lees, P. Stereo specific pharmacokinetics and pharmacodynamics of carprofen in the dog. J.Vet.Pharmacol.Ther. (1994), 17(6)447-54.

measurement of stereoselective metabolic and renal elimination after oral administration of carprofen in man. Drug. Metabol. And Disposition., (1989), 17, 414-419.

Meffin PJ, Sallustio B, Purdie Y. the specific binding of L-tryptophan to serum albumin. *Journal of Biological Chemistry 233*: 1436-1447, 1958.

Menzel, S., Waibel, R., Brune, K., Geisslinger, G.Is the formation of R-ibuprofenyl-adenylate the first stereo selective step of chiral inversion? Biochem Pharmaol 1994, 48: 1056-8.

Menzel-soglowek, S., Geisslinger, G., Beck, W.S. and Brune,K., Variability of inversion of (R) - flurbiprofen in different species., J.Pharm.Sci., (1992),81(9), 88-9.

Muller, S., Mayer, J.M., Etter, J.C., Testa, B.Influence of palmitate and benzoate on the unidirectional chiral inversion of ibuprofen in isolated rat hepatocytes. Biochem Pharmacol 1992, 44:1468-70.

Nagashina,H., Tanak, Y., Watanabe, H., Hayashi, R. and Kawada, K., Structural determination of Rat urinary metabolites of hoxoproflusodium, a new anti inflammatory Agent.Chem.Pharm.Bull., (1984), 32(1), 251-7.

Naruto,S., Tanaka, Y., Hayashi, R. and Terada,A., structural determination of rat Urinaryabolites of sodium 2-[4-(2-oxocyclo pentyl methyl) phenyl propionatedihydrate (loxoprofen sodium)a new anti inflammatory agent.. Chem.Pharm.Bull.,(1984),32(1), 258-67.

Nomura, T., Imai, T. and Otagiri, M., Biol.Pharm.Bull., (193), 16(3), 298-303.

nonsteroidal anti-inflammatory drugs-II.Racemization and hydrolysis of (R)- and (S) ibuprofen-coA thioesters. Biochem. Pharmacol., (1991), 42(10),1905-11.

Oliary,J., Tod,M., Nicolos, P., Petitjean, O. and Caille, G. Pharmacokinetics of ibuprofen enantiomers after single and repeatad doses in man. Biopharm. Drug. Dispos., (1992), 13(5), 337-44.

Oravcova,J., Mlynarik, V., Bysbicky,S., Soltes,L. and Sglay szalay, P., interaction of pirprofen enantionerery with human serum albumin Chirality., (1991),3, 412-17.

Otagri,M., Masuda,K., Imai,T.,Imamura,Y. and Yamasuki,M.,Binding of pirprofen to human serum albumin studied by dialysis and spectroscopy techniques Biochem. Pharmacol., (1989),38, 1-7.

Pedrazzini W., De angelis M., Zanoboni M.W. and Furgione A.Stereo chemical pharmacokinetics of the 2-aryl propionic acid ,nonsteriodal anti inflammatory drug Flunoxaprofen in rats and in man .Drug Research 38, 1170 (1988).

Perrin JH. A circular dichroic investigation of the binding of fenoprofen, 2(3-phenoxyphenyl)-propionic acid, to human serum albumin. *Journal of Pharmacy and Pharmacology 25*: 208-212, 1973.

Pliwal JK, Smith DE, Cox SR. *et al*. Stereo selective ,competitive and nonlinear Pasma protien binding of ibuprofen ernantiomers as determined invivo in healthy Subjectsj. Pharmacokinet Biopharm 1993;21: 145-61.

Reichel, C., Bang, H. Brune, K., Geisslinger, G., Menzel, S.2-Aryl propionyl co A epimerase:partial peptide sequence and tissue localization. J.Biol Chem. 1993 268: 3487-93.

Rendic S, Alebic-Kolbah T, Kajfez F, Sunjic V. Stereoselective binding of (+) and (-) α(benzoylphenyl) propionic acid (ketoprofen) to human serum albumin. Il Farmaco 35: 51-59, 1980.

Risdall PC.,AdamsS.S., Cramption,E.L. Marchant, B.The disposition and metabolism of flurbiprofen in several species including man. Xenobiotica., (1978), 8(11), 691-703.

Romero,A.J., Rackley. R.J, Rhodes.C.T., An evaluation of ibuprofen bionversion by simulation. Chirality, (1991), 3, 418-21.

Roy-De-Vos, M., Mayer, J.M., Etter, J.C., Testa, B.Clofibric acid increases the unidirectional chiral inversion of ibuprofen in rat liver preparations. Xenobiotica 1996, 26: 571-82.

Rubin A., Knadler M.P., HoP. P. K., Bechtol L.D. and Wolen R.L. Stereo selective inversion of (R) - fenoprofen to (S) - Fenoprofen in humans .. pharm. Sci. *74,* 82(1985).

Rubin,A., Chernish, S.M., Crabtree,R., Gruber, C.M., Warrick,P., Wolen,R.M. and Ridolfo, A.S., A profile of the physiological disposition and gastro intestinal effects of fenoprofen in man. Curr. Med.Res., (1974),2,529.

Rudy, A.C., Anliker, K.S., Hall, S.D.,High performance liquid chromatography determination of stereo isomeric metabolites of ibuprofen.J.Chromatogr 1990, 538: 395-405.

Rudy, A.C., Knight, P.M. Brate, D.C., Hall, S.D., Enantioselective disposition of ibuprofen in elderly persons with and without renal impairement.J. Pharmacol. Exp. Ther., 1995, 273: 88-93.

Rudy, A.C., Knight, P.M., Brate, D.C. hall, S.D.,Stereo selective metabolism of ibuprofen in humans:Administration of R-,S-,and racemic ibuprofen.J. Pharmacol. Exp. Ther., 1991, 259: 1133-9.

Rudy,A.C., Knight, P.M, Brater,D.C. and Hall,S.D., Stereo selective metabolism of ibuprofen in humans: administration of R -, S – and racemes

Runkel, R., Forchielli, E., Boost, G., Chaplin, M., Hill, R., Sevelius, H., Thompson, G. and Segre,EOpticalinversionof(2R)–to(2S) – isomers of Loxoprofen,anewantiinflammatory agent, and its monolydroxy metabolites in the Rat. J.Rheumatology., (1973), (Suppl, 2); 29-36.

Runkel, R.A., Chaplin, M., Boost, G., Segre, E. and Forchelli, E., J.Pharm.Sci., (1972), 703-8.

Sallustiro,B.C., Puridie, Y.J., White head, A.G., Ahern, M.J. and Meffin, P.T., The disposition of ketoprofen enantiomers in man. J.Clin. PHarmacol., (1988), 26, 765-770.

Scheuere, S., Oekers, R., Brune, K., Williams, K.M., Geisslinger, G. Naunyn-Schmied Effect of clofibrate on the chiral inversion of ibuprofen in healthy volunteers. Arch Pharmacol 1996, Abstract 23 R 147.

Shei WRand Chen, C.S., Purification and characterization of novel "2-Aryl propionyl-CoA epimerasis" from Rat liner Cytosol and Mitochondria. J.Biol.Chem., (1993)268, 3487-3493.

Shieh, W.R., Chen, C.S.,Purification and charactarisation of novel 2-aryl propionyl -co enzymeA epimerases from rat liver cytosaland mitochpndria.J. Biol Chem 1993, 268: 3487-93.

Siebler D, Kinawi A. Binding von razemischem Indoprofen undseiner Enantimeren an Humanserum-albumin> *Arzneimittel-Forschung 39*: 659-660, 1980.

Simmonds R.G., Woodage T. J., Duff S. M. and Green J. N.Stereo specific inversion of (R)-benoxaprofen in rat and man.Eur. J.Metab. Pharmacokin. *5,* 169 (1980).

Singh N. N., Jamali F., pasutto F. M., Russell A. S., Coutts R. T. and Darder K. S. Pharmacokinetics of enantiomers of Tia profenic acid in humans. *J. pharm. Sci. 75,* 439 (1986).

Sioufi A, Colussi D, Marfil F, Dubois JP. Determination of the (+) and (-) enantimoers of pirprofen in human plasma byhigh-performance liquid chromatography. *Journal of Chromatography 141*: 131-137, 1987.

Spahn H, Spahn I, Benet LZ .Probenicid induced changes in the clearance of carprofen enantiomers a preliminary study. Clin Pharmacol Ther 1989 May;45(5):500-5.

Spahn H, Spahn I, Pflugman G, Mutschler E. Measurement of carprofen enantiomer concentrations in plasma and urine using L-leucinamide as the coupling component. *Journal of Chromatography 433*: 331-338, 1988.

Stoltenborg JK, Puglisi CV, Rubio F, Vane FM. High-performance liquid chromatographic determination of stereoselective disposition of carprofen in humans. *Journal of Pharmaceutical Sciences 70*: 1207-1212, 1981.

Sudlow, G., Birkett, D.J. and Wade, D.N.,Further characterisation of Specific drug binding sites on human serum albumin. Mol. Pharmacol., (1976),12, 1052-61.

Suri A Grundy, B.L. and Derendorf, H. Pharmacokinetics and Pharmacodynamics of ibuprofen and flurbiprofen after oras administration. Int.J.Clin. Pharmacol. Ther., (1997)   35, 1-8.

Tan,S.C. *et al.*Ibuprofen stereochemistry double –the -trouble Enantiomer.,(1994), 4, 195-20.

Tanaka, Y., nishikawa, Y. and Hayashi, R., Species differences in metabolism of sodium 2-[4-{ 2-Oxo cyclopentylmethyl) – phenyl propionate dihydrate (

Loxoprofen sodium) a new anti inflammatory agent Chem.Pharm. Bull., (1983), 31, (10), 3656-64.

Thompson, G.F. and Collins,J.M. .Urinary metabolic profiles for choosing test animals for chronic toxicity studies ;application to naproxen. J.Pharm.Sci., (1973),62, 937.

Thompson, M.J., Rhys-williams, W., Lloyd, A.W. and Hanlon, G.W.,The stereo inversion of 2-aryl propionic acid nonmsteriodal anti inflammatory drugs and structurally related ccompound by verticillium lecanii. J.Apl.Microbiol., (1998), 85(1), 55-63.

Tracy, T.S. and Hall, S.D., Metabolic inversion of (R)- ibuprofen. Epimerization and hydrolysis of ibuprofenyl-CoA enzyme. Drug. Metab. Dispos., (1992), 20, 322-327.

Tracy, T.S., Hall, S.D.Determination of the epimeric composition of ibuprofenoy-coA. Anal Biochem 1991, 195: 24-9.

Tracy, T.S., Wirthwein, D.P., Hall, S.D.Metabolic inversion of R-ibuprofen.formation of ibuprofenyol –co enzyme A. Drug Metab Dispos 1993, 21: 114-20.

Verbeeck, K.R., Blackburn, L.J. and Loewen, R.G., Clinical pharmacokinetics of Non-steroidal Anti-inflammatory drugs Clin. Pharmacokinet., (1983),8,279-331.

Wagener HH. Kalbhen DA. Berner G. *et al*.Ibuprofen racemate and enantiomer.Akt Rheumatol 1991; 16: 65-9.

Walser S, Hruby R, Hesse E, Heinzl H, Mascher H. Arzni mittel for schceng 1997 Jun; 47 (6): 750-4.

Williams, K.M., Day, R.O., Knihinicki, R.D., Duffield, The stereo selective up take of ibuprofen enantiomers into adipose tissue. J.Lipid Res 1978, 19: 3-11.

Xiaotao, Q., Hall, S.D.Modulation of enantioselective metaboliusm and inversion of ibuprofen byxenobiotics in isolated rat hepatocytes. J.Pharmacol Exp Ther 1993, 266: 845-51.

Y. Nakamura, T. Yamaguchi, S. Takahashi, S. Hashimoto, K. Iwatani and Y. Nakagawa, J. Pharmacobio-Dynamics, 4, (1981), p. s-1.

Yamamoto, T., Mastura,K., Shintani,S., Hara,A., Miyabe, Y., Sugiyama, T. and Katagiri, Y.Dual effects of anti-inflammatory 2- aryl propionic acid derivatives in a major isoform of human liver 3-alpha hydroxyl steroid dehydroginase. Biol.Pharm. Bull., (1998), 21(1), 1148-53.

Young, M.A., Aaroms, L. and Toon, S., The pharmaco kinetics of the enantiomers of flurbi profen in patients with rheumatiod arthritis,J.Clin. Pharmacol., (1991), 31, 102-104.

Table I. Protein binding of enatimers (R,S) f nonsteroidal anti-inflammatory drugs

Drug	Protein and concentration	Concentration range	Tech	Binding	Model[a]	
					S	R